CHEMISTRY 02

INTRODUCTION TO CHEMISTRY LABORATORY MANUAL

BRONX COMMUNITY COLLEGE
DEPARTMENT OF CHEMISTRY & CHEMICAL TECHNOLOGY

Kendall Hunt
publishing company

Cover image © 2014, Shutterstock, Inc.

Kendall Hunt
publishing company

www.kendallhunt.com
Send all inquiries to:
4050 Westmark Drive
Dubuque, IA 52004-1840

ISBN 978-1-4652-4154-2

Printed in the United States of America

Contents

Acknowledgment

The laboratory experiments in this workbook are based on the original work of Professor Herman Stein. The experiments in their current state have been edited and submitted for publication by Professors Anthony Durante and John Molina.

Name: ______________________ Semester ______________ Date ______________

Course ______________ Section ______________ Instructor ______________________

LAB 1

Introduction to Lab Safety

Student Copy

For your protection you must obey these laboratory rules or forfeit the right to work in the lab!

1. **Familiarize** yourself with the location of the safety items in the laboratory: shower, eye-wash fountain, fire extinguisher and fire blanket. Make sure your pathway to them is never blocked.
2. Your work area (including the floor) should be absolutely clear. Remove all book-bags, books (except lab book and lab notebook) and briefcases from the lab bench area to the sides of the room.
3. **Wear provided, approved eye protection** (goggles) in the laboratory at all times. Goggles protect your eyes against both impact and splashes.
4. If you get a chemical in your eyes, wash them in the eye-wash fountain for 15 to 20 minutes and notify your instructor.
5. **Carefully follow all instructions,** both written and oral. Even "*safe*" experiments can be dangerous if instructions are not followed. If you do not know what to do, ask your instructor!
6. **No food or drink** is allowed in the laboratory. Never taste any chemicals in the laboratory. (Even common compounds like salt and sugar in the lab are not fit to eat.) Never use laboratory glassware to drink from.
7. **Smoking is prohibited** on campus and certainly not in Meister Hall, including the classrooms, labs and stairways. Violators will be reported to BCC Public Safety and may be debarred.
8. **Report all accidents** and glass breakages to your instructor **at once**. Ask to be taken to the infirmary for treatment of cuts, burns, or inhalation of vapors. Spilled mercury must be reported. Do not attempt to clean up broken glass, spilled reagents, or blood by yourself.
9. **Avoid breathing vapors** of any kind. Follow instructions for use of the ventilation hoods around the room. Usually chemicals stored in the hoods should not be removed but should be used there.
10. **Do not use mouth suction** to fill pipets in transferring chemical reagents. Use a rubber suction bulb. More injuries occur each year from mouth-pipetting than from any other error.

(over)

11. **Protect yourself!** Long hair must be confined with rubber bands or a hairnet when in the laboratory. Wear comfortable clothing in the lab. Shoes should have closed tops (no sandals). **A lab coat is strongly recommended.**

12. **Do not wear contact lenses** in the laboratory. They may absorb vapors which can permanently damage eyes.

13. **General Rules:**

    a. Never work in the laboratory without supervision.

    b. Never block the laboratory aisles.

    c. Never remove chemicals from the lab, and never attempt to do lab work at home.

    d. You should behave professionally and responsibly at all times.

By signing this document, I have read and understood the laboratory safety rules and agree to abide by them. I understand I may be removed from the lab room and may be debarred for failing to follow the rules.

Student name (print):____________________________ Student Signature:____________________

INTRODUCTION TO CHM 02

Equipment, Glassware and Lab Safety

Tongs, striker, test holder, spatula, dropper

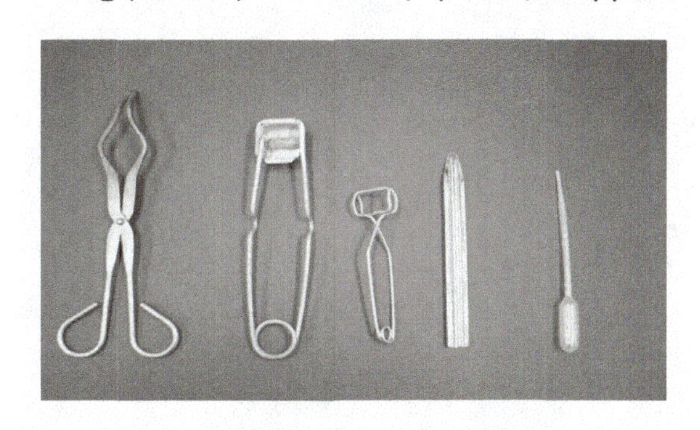

BEAKERS

GRADUATED CYLINDERS

ERLMEYER FLASKS

MULTIPLE BEAMS BALANCE

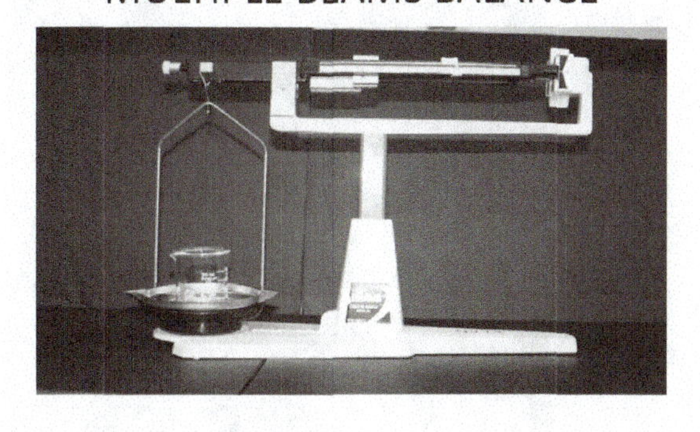

http://www.ohaus.com/input/tutorials/cog/COGentry.html

Bunsen Burner

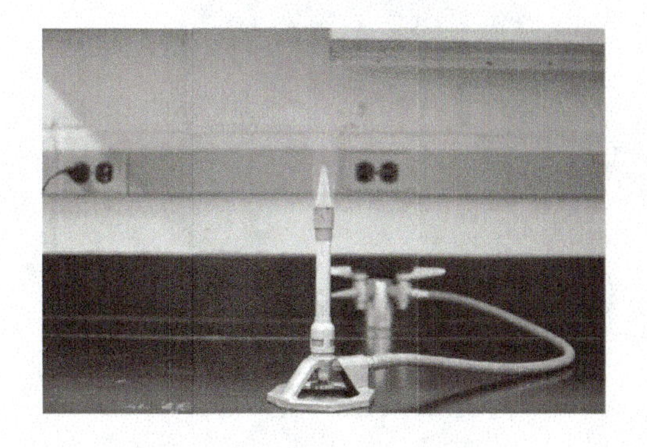

Thermometer

Name: ______________________ Semester ______________ Date ______________

Course ______________ Section ______________ Instructor ______________________

LAB 2
Physical and Chemical Change

Introduction and Discussion

Chemistry is the study of matter and the changes it undergoes. Changes in matter are divided into two categories: physical change and chemical change.

What are physical changes?

Physical changes do not involve a change in the composition of the substances in the sample. The appearance of the sample may change, but if the chemical composition remains unaltered, the change is described as physical. For example, if paper is shredded, the sample is still paper after shredding. If water freezes to ice the chemical composition of the substance doesn't change—it is still water. When sugar dissolves in water, the sugar and water coexist independently in the mixture, each retaining its own identity. These are all physical changes.

What are chemical changes?

On the other hand, chemical changes involve changes in the composition of the substance(s) being studied. The original materials—the *reactants*—are transformed into new substances, called *products,* whose compositions are different from the original materials. The characteristics of the reactants are completely lost, and new characteristics, belonging to the products, appear in their place.

How do we recognize chemical changes?

Since new substances are produced there are also changes in appearance (physical changes), which go along with chemical changes. These changes are often striking and unexpected, when compared to a situation where *only* physical changes occur. *We usually recognize a chemical change by observing physical changes in the products, which can only be attributed to the formation of new substances.*

When trying to decide if a change is *chemical,* ask yourself these questions:

1. Is there evidence that the original sample has been used up? If you have evidence that the reactants are no longer present after the change, then there has been a change in composition, and hence a chemical change has occurred.

2. Is there evidence that a new material has been formed? If you have evidence that a new material has formed, then the change is chemical.
3. Did you observe a spontaneous release of energy? Chemical changes frequently occur in nature with the release of energy in the form of heat or light.

Remember that the way you will judge that a reactant is used up is by the disappearance of the characteristics associated with that material. The way you tell a new product is formed is by the appearance of new characteristics *not* associated with the original reactant(s). If the change is chemical, you must make a judgment that the physical changes you see could only be the result of the original substance(s) being consumed and new substances being formed.

Experimental: Procedure and Record of Observations

1. Sample: Paper **Operation: Tearing**

   Examine a sample of paper and note (record) its physical properties. ____________

   ____

   ____

   Rip the piece of paper into several pieces.

   Re-examine the torn paper and note (record) its physical properties. ____________

   ____

   ____

   Has there been a substantial change in properties? ❒ Yes ❒ No

   Is the sample still paper after tearing? ❒ Yes ❒ No

   Do you have evidence for a change in composition? ❒ Yes ❒ No

2. Sample: Paper **Operation: Heating in air**

   Using tongs, set fire to a piece of paper by putting into a Bunsen burner flame.

   Examine the sample after burning and note (record) its physical properties. ____________

   ____

   ____

   Has there been a substantial change in properties? ❒ Yes ❒ No

   Is the sample still paper after burning? ❒ Yes ❒ No

   Do you think there has been a change in composition? ❒ Yes ❒ No

3. Sample: Copper **Operation: Heating in air**

   Examine a sample of copper wire and note (record) its physical properties. ____________

   ____

   ____

   Using tongs, heat the copper wire in a flame for a full three minutes. Be sure to heat the wire in the hottest part of the flame (ask your instructor to demonstrate this).

   Examine the sample of copper wire after heating and note (record) its physical properties.

   ____

   ____

   Has there been a substantial change in properties? ❒ Yes ❒ No

   Can the black coating be scraped off? ❒ Yes ❒ No

   Does the black coating have the same properties as copper? ❒ Yes ❒ No

   Is the black coating a new substance? ❒ Yes ❒ No

   Do you think there has been a change in composition, at least on the surface of the copper wire? ❒ Yes ❒ No

4. Sample: Magnesium wire **Operation: Heating in air**

   Examine a sample of magnesium wire and note (record) its physical properties. ____________

   ____

   ____

   Caution: Do not look directly at the magnesium wire when you heat it in the next step.

   Using tongs, heat the magnesium wire in a flame for a couple of minutes.

   Examine the residue left from the magnesium wire after heating and note (record) its physical properties.

   ____

   ____

Has there been a substantial change in properties?	❐ Yes	❐ No
Was there an obvious release of energy?	❐ Yes	❐ No
Do you think there has been a change in composition?	❐ Yes	❐ No
Do you think the white ash is a new substance?	❐ Yes	❐ No

5. Sample: Nichrome wire **Operation: Heating in air**

   Examine a sample of nichrome wire and note (record) its physical properties. ____________

   ____

   ____

   Using tongs, heat the nichrome wire in a flame for three minutes.

   Examine the sample of nichrome wire after heating and note (record) its physical properties.

   ____

   ____

Has there been a substantial change in properties?	❐ Yes	❐ No
Can the dark coating be scraped off?	❐ Yes	❐ No
Is there evidence for a new substance?	❐ Yes	❐ No
Do you think there has been a change in composition, at least on the surface of the nichrome wire ?	❐ Yes	❐ No

6. Sample: Iodine crystals **Operation: Heating in air**

Examine some iodine crystals and note (record) their physical properties. (Do not open sample jar). ______________________________

____

____

DEMO: The instructor will place a few iodine crystals in a dry, clean test tube. Using a test tube holder (not tongs), the sample will be gently warmed over a flame carefully so as to avoid allowing fumes to escape from the tube. **(Breathing iodine vapors is dangerous. Demo is best done in the hood.)**

Describe the changes that occur. ______________________________

____

____

Is there evidence for a new substance?	❒ Yes	❒ No
Is the purple vapor a new substance?	❒ Yes	❒ No
Is it possible that this purple gas is iodine vapor?	❒ Yes	❒ No
Is the shiny material on the inside walls of the tube a new substance?	❒ Yes	❒ No
Is it possible that the shiny material is iodine?	❒ Yes	❒ No
Do you think there has been a change in composition?	❒ Yes	❒ No

7. Sample: Iodine crystals **Operation: Addition of alcohol**

DEMO: The instructor will place an iodine crystal in a dry, clean test tube, and then add about 1 to 2 mL of ethyl alcohol.

Describe the changes that occur. ______________________________

____

____

Is the brown liquid a new substance?	❒ Yes	❒ No
Is it possible that the iodine is dissolved in the alcohol?	❒ Yes	❒ No
Is there evidence for a new substance?	❒ Yes	❒ No
Do you think there has been a change in composition?	❒ Yes	❒ No

8. Sample: Sodium **Operation: Addition to water**

   Your instructor will cut a small chip of sodium metal from a larger piece of sodium metal.

   Examine the sodium chip and record its physical properties. ______________________

   ____

   ____

   DEMO: The instructor will add the sodium to about 100 mL of water in a 400 mL beaker.

   Describe the changes that occur. ______________________

   ____

   ____

   Was the reactant used up? ❐ Yes ❐ No

   Is there evidence for a new substance? ❐ Yes ❐ No

   Do you think there has been a change in composition? ❐ Yes ❐ No

9. Sample: Zinc **Operation: Addition to hydrochloric acid**

   Your instructor will take a small sample of zinc flakes or powder from a container.

   Examine the zinc sample and record its physical properties. ______________________

   ____

   ____

   DEMO: The instructor will add the zinc to about 5 mL of hydrochloric acid in a 6-inch test tube.

   Describe the changes that occur. ______________________

   ____

   ____

   Was the reactant used up? ❐ Yes ❐ No

   Is there evidence for a new substance? ❐ Yes ❐ No

   Do you think there has been a change in composition? ❐ Yes ❐ No

10. Sample: Sodium carbonate **Operation: Addition to hydrochloric acid**

Your instructor will take a small sample of sodium carbonate powder from a container.

Examine the sample of sodium carbonate and record its physical properties. ______________

____

____

DEMO: The instructor will add the sodium carbonate to about 5 mL of hydrochloric acid in a 6-inch test tube.

Describe the changes that occur. ______________________________

____

____

Is there evidence for a new substance? ❐ Yes ❐ No

Do you think there has been a change in composition? ❐ Yes ❐ No

11. Sample: Sodium chromate and lead nitrate solutions **Operation: Mixing**

Examine the solutions and record their physical appearance. ______________

____

____

DEMO: The instructor will mix 1 or 2 mL of each of the solutions in a small test tube.

Describe the changes that occur. ______________________________

____

____

Is there evidence for a new substance? ❐ Yes ❐ No

Do you think there has been a change in composition? ❐ Yes ❐ No

12. Sample: Sulfur **Operation: Heating in air**

Your instructor will take a small sample of sulfur powder from a container.

Examine the sample of sulfur and record its physical appearance. ____________________

____

____

DEMO: The instructor will burn the sample of sulfur on a spatula in the hood.

Describe the changes that occur. ____________________

____

____

Was the reactant used up?	❒ Yes	❒ No
Is there evidence for a new substance?	❒ Yes	❒ No
Do you think there has been a change in composition?	❒ Yes	❒ No

Data Tables

Summarize your observations by completing the two data tables on pages 13 and 14.

Conclusions

Make a judgment about whether the change was physical or chemical in the section entitled Conclusions.

Supplemental Questions

Fill in the answers to the questions on the last page.

Submit the report to your instructor at the next meeting of lab.

Data Table

Physical and Chemical Change

Reactants	Properties of Reactants	Operation	Observations	Properties of Products	New Substance Formed? (Yes/No)
Paper		tearing			
Paper and air		heating			
Copper and air		heating			
Magnesium and air		heating			
Nichrome and air		heating			
Iodine and air		heating			
Iodine and alcohol		adding			

Reactants	Properties of Reactants	Operation	Observations	Properties of Products	New Substance Formed? (Yes/No)
Sodium and water		adding			
Zinc and hydrochloric acid		adding			
Sodium carbonate and hydrochloric acid		adding			
Lead nitrate and sodium chro-mate		adding			
Sulfur and air		heating			

Conclusions

Make your conclusions on the tests performed by completing the table below.

Reactants	Operation	Type of Change (Physical or Chemical)
Paper	tearing	
Paper and air	heating	
Copper and air	heating	
Magnesium and air	heating	
Nichrome and air	heating	
Iodine and air	heating	
Iodine and alcohol	adding	
Sodium and water	adding	
Zinc and hydrochloric acid	adding	
Sodium carbonate and hydrochloric acid	adding	
Lead nitrate and sodium chromate	adding	
Sulfur and air	heating	

Supplemental Questions (Answer these questions as part of your report.)

1. When no new substance forms, a ________________ change occurs.
2. The substance present in air that is required for the chemical changes that occurred when air was involved in the above tests is ________________ .
3. The process whereby a sample, such as iodine, changes directly from solid to gas is called ________________________ .
4. Classify each of the following as a physical or chemical change:

soda pop fizzes	❒ Physical	❒ Chemical	(Check ✔ one)
banana slices turn dark	❒ Physical	❒ Chemical	
dry ice sublimes	❒ Physical	❒ Chemical	
milk goes sour	❒ Physical	❒ Chemical	
(polluted) rain turns acidic	❒ Physical	❒ Chemical	
flashlight batteries discharge	❒ Physical	❒ Chemical	
iron rusts	❒ Physical	❒ Chemical	
sugar dissolves in warm water	❒ Physical	❒ Chemical	

Name: ______________________	Semester ______________	Date ______________
Course ______________ Section ______________	Instructor ______________________	

LAB 3

Elements and Compounds

Introduction

It is important to know what properties are used to distinguish metallic elements from nonmetallic elements. It is equally important to be able to express correlations between these properties and the type of element (metal or non-metal) which exhibits them.

In this experiment, we will examine actual samples of elements, determine some of their properties, and try to relate their properties to the type of the element.

Principles Involved

Elements are divided into two major types: metals and non-metals.

An element is classified as a metal if, in general, it has most of the physical and chemical properties associated with the class called metals. Among the typical *physical* properties of metals are luster, good thermal and electrical conductivity, ductile, maleable, relatively high densities, high melting points, and high boiling points. Among the typical *chemical* properties of metals is their tendency to combine with non-metals, such as, for example, oxygen, sulfur and chlorine.

Elements which do not have these properties (or have them to a minor degree) are classified as nonmetals. Individual elements may exhibit exceptional properties in some cases. It is important to realize that classification as metal or non-metal is a generalization, and that exceptions do exist.

Procedure

PART ONE: Physical Properties ***(complete the data table on page 19)***

1. Inspect the samples of the eleven elements provided by your instructor. Note and record the physical properties (homogeneity, state, luster, color) of each.
2. Your instructor will demonstrate the use of a volt meter (conductivity meter) to determine the electrical conductivity of **aluminum, carbon, copper, iron, magnesium, sulfur, tin,** and **mercury.**
3. Determine the thermal conductivity of **aluminum, copper,** and **iron** by holding a piece of the wire with your fingers and *cautiously* warming it in a low flame.
4. Physical constants (density, melting point, and boiling point) of the elements are provided. Compare these values to form generalities.

PART TWO: Chemical Properties ***(complete the data table on page 20)***

A decomposition reaction will be demonstrated by your instructor.

1. The three combination reactions illustrate the tendency of metals to combine with non-metals to form binary ionic compounds.
2. A binary ionic compound will be decomposed to show that it is composed of two elements (a metal and a non-metal).

Data

Record the data in the tables on pages 19 and 20.

Conclusions

Follow directions on page 21.

Data Table Physical Properties

Elements and Compounds

Elements							Conductivity				
Name	**Symbol**	***metal or non-metal***	**Homogeneous (yes/no)**	**State**	**Luster**	**Color**	**Elect**	**Heat**	**Density g/mL**	**M.P. °C**	**B.P.° C.**
Aluminum									2.7	660	2246
Copper									8.92	1083	2595
Magnesium									1.74	651	1107
Mercury							DEMO		13.6	-39	356
Iron									7.86	1535	3000
Tin									7.28	232	2260
Bromine									2.92	–7	59
Carbon									3.5	>355	4827
Chlorine									0.003	–100	–34
Phosphorus									1.82	44	280
Sulfur									2.0	112	445

Data Table Physical Properties

Elements and Compounds

Combination Reactions A + B → AB

WORDS EQUATIONS **METALS combine with NON-METALS to produce BINARY IONIC COMPOUNDS**	**Evidence for Reaction (Observations)**
Magnesium + oxygen → Magnesium oxide	
Copper + oxygen →	
Iron + oxygen →	

Try naming the compounds formed. Enter names in the data table.

DECOMPOSITION REACTION AB → A + B **(Demo by the Instructor)**

BINARY COMPOUNDS can be decomposed into TWO ELEMENTS	**Evidence for Reaction (Observations)**
Mercury(II) oxide + Heat → ____________ + ____________ + ____________	

Enter the names of the elements or compounds formed.

Conclusions (based on your observations during this experiment)

1. Which of the physical properties recorded can be used to distinguish metals from non-metals? Can color, homogeneity be used to distinguish metals from non-metals?

2. Do your observations support the statement that luster is a typical physical property of metals? Are there exceptions?

3. Do your observations support a generalization that metals are solids? Are there exceptions?

4. Do your observations support a statement that electrical conductivity is a typical physical property of metals? Are there exceptions?

5. Do your observations support a statement that metals have relatively high densities? Are there exceptions?

6. Do your observations support a generalization that metals have relatively high melting points and boiling points? Are there any exceptions?

7. Do your observations support the conclusion that carbon is a non-metal? How do you reconcile your observations about carbon with the fact that carbon is a non-metal? (Attach additional pages, as needed.)

Name: ______________________________	Semester ______________	Date ______________
Course ______________ Section ______________	Instructor ______________________	

LAB 4

Techniques of Measurement

Assigned Color (circle one): **Blue** **Red** **Yellow** **White**

This laboratory exercise is a self-test to determine if you are able to read scales occurring on common types of laboratory equipment. You should complete the preliminary exercises on how to read scales, at the following web-link: *http://us.ohaus.com/us/en/home/markets/education-world/teachers-and-schools/balance-tutorials.aspx*. Select the balance tutorials and practice weighings for the OHAUS Triple beam and the Model 311 Cent-o-gram, complete the tutorials and the exercises provided.

There are ten "Stations" in the laboratory. At nine of the stations you are asked to make measurements. The samples at each station are color-coded. From the four colors above, circle the color you are assigned to work with.

After making a measurement, record it on this sheet in the space provided. When you have completed all these stations, check your answers with the "correct" values at the Instructor's desk. For any answers which do not agree with the "correct" values, return to that station and redo your work to see where you might have been mistaken. If you cannot find your mistake, see your Instructor for help.

Except for Station 1, remember to make each measurement as precisely as possible. *Record each number you are certain of, plus one estimated digit.* Try to do your own work, as you will need to be able to use laboratory equipment correctly and independently if you are to succeed in any science.

****

STATION 1

Use the ruler to measure the item to the nearest inch.
(Round off your answer to the nearest whole number at this station only.)

Length of a test tube __________ in

For the rest of your measurements, be as precise as you can! Do not round off.

STATION 2

Use the ruler to measure the cylinder in centimeters.

Length of brass cylinder __________ cm

STATION 3

Use the ruler to measure the block in centimeters.

A. Length of the brass block __________ cm

B. Width of the brass block __________ cm

STATION 4

Use the meter stick to measure each item in meters.

A. Height of the fire blanket cabinet __________ m

B. Width of the fire blanket cabinet __________ m

STATION 5

Read the volume of liquid in each graduated cylinder.

A. 10 mL cylinder __________ mL

B. 25 mL cylinder __________ mL

C. 50 mL cylinder __________ mL

D. 1 liter cylinder __________ mL

STATION 6

Read the burette at the level of liquid.

A. Burette reading __________ mL

STATION 7

Record total weight added to the beams on each balance.

A. Triple beam balance __________ g

B. Triple beam balance __________ g

STATION 8

Record total weight added to the beams on each balance.

A. Multiple beam balance __________ g

B. Multiple beam balance __________ g

STATION 9

Record the number stamped on the cylinder. Weigh the cylinder on the multiple beam balance. Record the weight.

Cylinder No. _____ __________ g

Compare your measurements with the values at the Instructor's desk. Return to any station where there is an error and re-examine the scale.

STATION 10

CALCULATIONS: (You may finish this section at home if you do not have enough time in lab) Perform the following calculations using the factor-label method.

Use the corrected results obtained at the above stations for the original measurement in each case. Show the set-up for each conversion.

Station 1. Convert the length from in to cm. __________ cm

Station 2. Convert the length from cm to in. __________ in

Station 3B. Convert the width from cm to mm. __________ mm

Station 4A. Convert the height from m to cm. __________ cm

Station 4B. Convert the width from m to km. __________ km

Station 5A. Convert the volume from mL to L. __________ L

Station 5B. Convert the volume from mL to μL. __________ μL

Station 5D. Convert the volume from mL to L. __________ L

Station 8A. Convert the weight from g to mg. __________ mg

Station 8B. Convert the weight from g to kg. __________ kg

Name: ____________________	Semester ____________	Date ____________
Course ____________ Section ____________	Instructor ____________	

LAB 5

Density of Liquids

Objectives

- To reinforce scale reading skills using common lab equipment (graduated cylinders, balances).
- To teach the technique of weighing a sample by difference.
- To illustrate a laboratory method of determining the density of a liquid sample.

Introduction and Discussion

What is density? ____________________

Density is defined as a ratio of the mass of a sample to its volume.

$$\text{Density} = \frac{\text{mass of sample}}{\text{volume of sample}}$$

What information does density give us? ____________________

Density tells us how closely matter is packed in the sample. For example, the density of water is 1.00 g/mL and the density* of ethyl alcohol is 0.789 g/mL. This means for water, an amount of matter equal to 1.00 gram is packed into one milliliter. For ethyl alcohol, only 0.789 gram of matter is in one milliliter. Water is more dense.

If the sample is a pure substance, the density is a *characteristic* physical property of that substance. It is rare that any two substances will have identical densities. Therefore, density finds use in the identification of substances. If a pure sample is found to have a density* of 0.789 g/mL, it cannot be water, and there is a high probability that it is ethyl alcohol.

*measured at 20°C

What is the unit of density? ____________________

In the metric system, the units used to measure the density of liquid and solids are grams per milliliter (g/mL,) or grams per cubic centimeter (g/cc = g/cm^3).

There are many different ways to experimentally determine the density of a liquid. Two different methods are described here.

Experimental Procedure

Work through the following procedure twice, first using water as the sample, and then using an unknown liquid as the sample.

Record all measurements in the data section (on page 29).
Note that the graduated cylinder serves two purposes in the procedure used:

(1) as a container to conveniently weigh the liquid samples.
(2) as a device for measuring volume.

1. Weigh a dry, empty 10 mL graduated cylinder on a Cent-o-gram balance. Record the weight as accurately as possible.
   (The last number should be estimated, giving a reading with three digits after the decimal point.)

2. Pour between 6 and 9 mL of the sample into the cylinder.

3. Weigh the cylinder with the sample in it. Record the weight as accurately as possible.
   (The last number should be estimated, giving a reading with three digits after the decimal point.)

4. Record the volume in the cylinder as accurately as possible.
   (The last number should be estimated, giving a reading with two digits after the decimal point.)

5. Record the temperature as accurately as possible. *(One digit after the decimal point.)*

6. Calculate the weight of the sample. *(Three digits after the decimal point.)*

7. Calculate the density of the sample. *(Two digits after the decimal point.)*

Data Tables

- Enter measured and calculated values in the tables below.
- When you record data, be sure the last digit recorded is an estimated number.
- Show Table I to the instructor for approval *before* going on to the unknown sample.
- The units in Table I are given. You must record the number with units in Table II.

TABLE I	Sample: water.	Temperature of water _____ °C
	Weight of the empty cylinder	__________ g
	Weight of cylinder + water	__________ g
	Weight of water	__________ g
	Volume of water	__________ mL
	Density of water	__________ g/mL
		Instructor's approval: _____ (initials)

TABLE II	Sample: Unknown	Temperature of water _____ °C
	Weight of the empty cylinder	__________ g
	Weight of cylinder + unknown	__________ g
	Weight of unknown	__________ g
	Volume of unknown	__________ mL
	Density of unknown	__________ g/mL
		Instructor's approval: _____ (initials)

Calculations

Show how you calculated the weight of the water and the density of the water.

Show how you calculated the weight of the unknown and the density of the unknown.

Conclusions

According to the data in this experiment:

Your calculated density of water was determined to be: __________ g/mL

Your calculated density of the unknown was determined to be: __________ g/mL

Supplemental Questions

1. Why is it important to record the temperature at which the density of a sample is measured?

2. List the names of some calibrated laboratory glassware that can be used to measure the volume of liquids.

3. Consider two samples of liquids:

   Sample (A): 3.48 mL of mercury (d = 13.6 g/mL)

   Sample (B): 60.0 mL of alcohol (d = 0.789 g/mL)

   Which sample has the greatest volume?
   (Check answer. Keep 3 digits when you compare numbers.)

   ❑ Sample A ❑ Sample B ❑ Same

   Which sample has the greatest density?

   ❑ Sample A ❑ Sample B ❑ Same

   Which sample has the greatest mass?

   ❑ Sample A ❑ Sample B ❑ Same

4. Will a 500-gram sample of carbon tetrachloride have a greater density than a 7.0-gram sample of carbon tetrachloride?

   Explain your answer. _____Yes _____No

Name: ______________________	Semester ______________	Date ______________
Course ______________ Section ______________	Instructor ______________________	

LAB 6

Density of Solids

Objectives

- To reinforce scale reading skills using common lab equipment (metric rulers, graduated cylinders, balances).
- To teach the technique of weighing a sample by difference.
- To illustrate two different methods of determining the density of solids.

Introduction and Discussion

What is density? ______________________

Density is defined as a ratio of the mass of a sample to its volume.

$$\text{Density} = \text{Mass} / \text{Volume}$$

What information does density give us? ______________________

Density tells us how closely matter is packed in the sample. For example, the density of silver is 10.5 g/mL and the density of magnesium is 1.75 g/mL. This means for silver, an amount of matter equal to 10.5 grams is packed into one milliliter. For magnesium, only 1.75 grams of matter is in one milliliter. Silver is more dense.

If the sample is a pure substance, the density is a characteristic physical property of that substance. It is rare that any two substances will have identical densities. Therefore density finds use in the identification of substances. If a pure sample is found to have a density* of 10.5 g/mL, it cannot be magnesium, and there is a high probability that it is silver.

*measured at 20°C

How is density measured? ______________________

In the metric system, the units used to measure the density of liquid and solids are grams per milliliter (g/mL) or grams per cubic centimeter (g/cc or g/cm^3).

There are many different ways to experimentally determine the density of solid samples. Two methods are described in the experimental section which follows.

Experimental Procedure

Part 1: The Density of a Regular Solid

Regular solids have a volume that can be calculated using formulae from geometry.

Examples:	Cube's	Volume = (length) × (width) × (height)	
	Cylinder's	Volume = $\pi \times (\text{radius})^2 \times (\text{height})$	(π = 3.14)

1. Weigh the metal cylinder on a Cent-o-gram balance. Record the weight as accurately as possible.
   (Estimate the last number to report three digits after the decimal.)

2. Using the metric ruler provided, measure the height of the cylinder in centimeters.
   (Estimate the last number to report two digits after the decimal.)

3. Using the ruler, measure the diameter of the cylinder in centimeters.
   (Estimate the last number to report two digits after the decimal.)

4. Calculate the radius of the cylinder. *(Radius = diameter/2)*

5 Calculate the volume of the cylinder *(Volume = $\pi r^2 h$)*

6. Calculate the density of the cylinder. *(Density = mass/volume; report the density to two digits after the decimal).*

Data Tables

- Enter measured and calculated values in the table below.
- Be sure that proper units are recorded for each value.
- When you record data, be sure the last digit recorded is an estimated number.
- Show Table I to the instructor for approval before going on to the next sample.

TABLE I **Sample: metal cylinder** **Room temperature ____ °C**

Sample: metal cylinder	
Weight of metal cylinder	________
Height of cylinder	________
Diameter of cylinder	________
Radius of cylinder*	________
Volume of cylinder*	________
Density of cylinder*	________

* Calculated values. (see Calculations section below)

Have you recorded the proper units next to each number?

Instructor's approval: ____ (initials)

Calculations

*Show how you calculated the radius, the volume, and the density of the metal cylinder.

Part 2: Density of an Irregular Solid

The shape of some solids does not permit the use of formulas to calculate their volume. Other techniques must be used in these cases. If the sample is heavier than water and does not dissolve in water, the volume of the sample can be determined "by displacement," i.e., the volume of the sample is equal to the volume of the water it displaces when it sinks. This is the concept behind the following procedure to determine the density of lead shot.

1. Obtain a 10.00 mL graduated cylinder. Place approximately 5 mL of water in the cylinder.

   Record the exact volume of the water in the cylinder. *(Estimate the last number; report two digits after the decimal.)*

2. Weigh the graduated cylinder, with the water in it, on a Cent-o-gram balance.

   Record the weight as accurately as possible. *(Estimate the last number, three digits after the decimal.)*

3. Obtain sonic lead shot. Carefully pour the shot into the cylinder so that the water level rises between 3 to 4 mL.

   Record the exact volume occupied by both the water and the lead. *(Estimate the last number, to report a reading with two digits after the decimal point.)*

4. Weigh the graduated cylinder, with the water and the lead in it, on a Cent-o-gram balance.

   Record the weight as accurately as possible. *(Estimate the last number, three digits after the decimal.)*

5. Calculate the volume occupied by the lead.

6. Calculate the weight of the lead.

7 Calculate the density of the lead. Report the density to two digits after the decimal.

Data Tables

- Enter measured and calculated values in the table below.
- Be sure that proper units are recorded for each value.
- When you record data, be sure the last digit recorded is an estimated number.

TABLE I **Sample: lead shot** **Room temperature ____ °C**

Weight of cylinder + water	________
Weight of cylinder + water + lead	________
Weight of lead*	________
Volume of water	________
Volume of water + lead	________
Volume of lead*	________
Density of lead*	________

Instructor's approval: ____ (initials)

Calculations

*Show calculations for the weight of the lead, the volume of the lead, and the density of the lead.

Conclusions

According to the data in this experiment:

Your density of the metal cylinder was determined to be: ________ g/mL

Your density of the lead shot was determined to be: ________ g/mL

Supplemental Questions

1. Which of the following metals is most likely the metal in the cylinder you used in Part I?

	❏ Magnesium	❏ Aluminum	❏ Zinc	❏ Iron	❏ Silver
density =	1.75 g/cm^3	2.70 g/cm^3	7.10 g/cm^3	7.85 g/cm^3	10.5 g/cm^3

2. If the cylinder you used in Part I were pure gold, how much would it weigh? *(The density of gold is 19.3 g/cm^3.)*

3. Consider samples of two different metals:

Sample (A): 60.0 mL of magnesium (d = 1.75 g/mL)

Sample (B): 10.0 mL of silver (d = 10.5 g/mL)

Which sample has the greatest volume?
(Check answer. Keep three digits when you compare numbers.)

❏ Sample A ❏ Sample B ❏ Same

Which sample has the greatest density?

❏ Sample A ❏ Sample B ❏ Same

Which sample has the greatest mass?

❏ Sample A ❏ Sample B ❏ Same

4. Express the ratio of the density of silver (d = 10.5 g/mL) to the density of magnesium (d = 1.75 g/mL) in terms of a whole number ratio.

Answer: The ratio is _____ to _____ .

Name: ______________________ Semester ______________ Date ______________

Course ______________ Section ______________ Instructor ______________________

LAB 7

Types of Chemical Reactions

Objectives

- To illustrate four general types of chemical reactions.
- To review chemical nomenclature.
- To reinforce formula writing skills.
- To practice balancing equations.

Introduction and Pre-lab Discussion

I. The Charges on Ions

A. Simple Ions (ions formed from only one atom)

Ions form from atoms which gain or lose electrons. Metals tend to lose electrons to form positive ions. Non-metals tend to gain electrons to form negative ions. It is sometimes very easy to predict the charges on simple ions from their position in the Periodic Table. You can relate the charge on each of the following simple ions to the location in the Periodic Table of the element from which it forms.

Group IA		Group IIA		Group IIIA		Group VIA		Group VIIA	
Hydrogen	H^+							Hydride	H^-
Lithium	Li^+					Oxide	O^{2-}	Fluoride	F^-
Sodium	Na^+	Magnesium	Mg^{2+}	Aluminum	$A1^{3+}$	Sulfide	S^{2-}	Chloride	Cl^-
Potassium	K^+	Calcium	Ca^{2+}					Bromide	Br^-
		Strontium	Sr^{2+}					Iodide	I^-
		Barium	Ba^{2+}						

Note that the name of the positive ion has the same name as the metal from which it formed.
Note that the name of the negative ion is derived from the name of the non-metal from which it formed by adding an *-ide* suffix.

With the Group B Elements (the Transition Elements), correlation of the charge on the metal ion and its position in the Periodic Table is not so obvious. Furthermore, some transition elements have more than one common ion. All of the Transition Elements are metals and therefore will form positive ions. Some important ions formed by transition elements are:

Silver	Ag^{+}	Iron(II)	Fe^{2+}	Iron(III)	Fe^{3+}		
Copper(I)	Cu^{+}	Copper(II)	Cu^{2+}				
		Tin(II)	Sn^{2+}			Tin(IV)	Sn^{4+}
		Lead(II)	Pb^{2+}			Lead(IV)	Pb^{4+}
		Zinc	Zn^{2+}				
Mercury(I)	Hg^{+}	Mercury(II)	Hg^{2+}				

Note that many Transition Elements form ions with a 2+ charge.

When the same metal forms two ions, the ions are named either by putting the magnitude of the charge in Roman numbers after the name of the metal, or by using suffixes on the Latin stem of the word for the element. The suffix *-ous* is used for the lower of the two charges, and the suffix *-ic* is used for the higher value. Thus *ferric* means the same as iron(III), cuprous means the same as copper(I), etc.

B. Polyatomic ions or Radicals

These are groups of non-metal atoms bearing an overall charge. A polyatomic ion functions as a unit in chemical reactions. Their formulas and charges must be memorized.

There is only one important positive radical, the ammonium ion, NH_4^{+}. All other important radicals are negatively charged.

Nitrate	NO_3^{-}	Sulfate	SO_4^{2-}	Phosphate	PO_4^{3-}
Nitrite	NO_2^{-}	Sulfite	SO_3^{2-}		
Hydroxide	OH^{-}	Carbonate	CO_3^{2-}		
Cyanide	CN^{-}	Chromate	CrO_4^{2-}		

II. Formula Writing

Ionic compounds are electrically neutral, even though they contain charged particles. This means that the sum of all the positive charges in a compound must equal the sum of all the negative charges, so that the total charge on the compound is zero. The formula of the compound indicates, by the use of subscripts, the simplest ratio of ions necessary so positive charges and negative charges cancel.

Examples:

One Na^+ ion requires one Cl^- ion to cancel charges. The formula for sodium chloride is $NaCl$.
One Ca^{2+} ion requires two Cl^- ions to cancel charges. The formula for calcium chloride is $CaCl_2$.
Two Na^+ ions are required for one O^{2-} ion. The formula for sodium oxide is Na_2O.
One Al^{3+} ion requires three OH^- ions to cancel charges. Aluminum hydroxide is $Al(OH)_3$.

III. Pre-Lab Exercise

Pre-Lab Exercise: Enter the formulas for the following compounds. Refer to the lists of ions above if necessary.

sodium hydroxide	______	ammonium chloride	______	sodium carbonate	______
silver nitrate	______	potassium chromate	______	barium sulfate	______
lead(II) iodide	______	ferric hydroxide	______	sodium sulfate	______

IV. Predicting Products

It is important to understand that simply because a hypothetical reaction can be described in terms of a chemical equation on paper or on the blackboard does not necessarily mean it will actually occur in the laboratory. In the laboratory exercise which follows be sure to indicate the observed evidence that indicates a chemical change is occurring.

There are instances when we can actually predict with assurance that a reaction will occur.

1. In double displacement reactions between acids and bases, the formation of water from H^+ and OH^- is expected for reaction, but is often not evident. However, the formation of water can be predicted from writing the chemical equation.
2. In double displacement reactions between salts, the formation of a precipitate is evidence for a reaction. Precipitate formation can be predicted using a solubility table.
3. In single displacement reactions, a reaction will occur if the reacting element is more active than the element it displaces. This can be predicted using the Activity Series.

V. Balancing Equations

All chemical equations must be balanced. Only balanced equations conform to the Law of Conservation of Mass. Before attempting to balance, the formulas of the substances involved must be correctly written. Balancing the equations is accomplished by placing a coefficient before the formulas of the substances so that the total number of atoms of each element is the same on each side of the equation. Subscripts in the formulas cannot be altered.

Unbalanced: $Al(OH)_3 + H_2SO_4 \rightarrow Al_2(SO_4)_3 + H_2O$

Balanced: $2\ Al(OH)_3 + 3\ H_2SO_4 \rightarrow Al_2(SO_4)_3 + 6\ H_2O$

Experimental

The reactions to be carried out in the lab are identified in the Data Tables in the Report section. Some of the reactions (those which are dangerous) will be demonstrated by your instructor. The remaining reactions will be done by the student using the reagents in the chemical kits provided.

Use only 1 mL of the reagents (which are in aqueous solution) for your tests.

Start the Laboratory Tests and complete the Data Tables.

VI. Reaction Types

Four types of reactions are illustrated in this laboratory exercise. You should recognize the distinctive features of each type:

1. Combination Reactions: Two reactants form one product.

In general,	Reactant A	+	Reactant B	→	Product C
Examples:	C	+	O_2	→	CO_2
	SO_3	+	H_2O	→	H_2SO_4

2. Decomposition Reactions: One reactant forms two (or more) products.

In general,	Reactant A	→	Product B	+	Product C
Examples:	H_2CO_3	→	CO_2	+	H_2O
	2HgO	→	2Hg	+	O_2

3. Single Displacement Reactions: An element reacts with a compound to produce a new element and a new compound.

In general,	Element M	+	Reactant A-B	→	Product M	+	New Element
Examples:	Mg	+	H_2SO_4	→	$MgSO_4$	+	H_2
	Zn	+	$CuSO_4$	→	$ZnSO_4$	+	Cu

4. Double Displacement Reactions: Two compounds react to form two new compounds by re-pairing of ions.

In general	Reactant A-B	+	Reactant C-D	→	Product A-D	+	Product C-B
Examples:	$AgNO_3$	+	NaCI	→	$NaNO_3$	+	AgCl
	KOH	+	HNO_3	→	KNO_3	+	H_2O

Bronx Community College/CUNY
Department of Chemistry & Chemical Technology

*Report Form: **Types of Chemical Reactions***

Report Form:

Experiment # 7: TYPES OF CHEMICAL REACTIONS

Complete the following tables for the chemical reactions you have studied in the laboratory by entering the *formulas* for the reactants and the products. After the correct formulas have been entered, *balance* each equation. Submit report form to instructor.

Student's Name: ______________________________

Course: CHM02 Section: ______________________________

Instructor's Name: ______________________________

Date: ______________________________

Bronx Community College/CUNY
Department of Chemistry & Chemical Technology

Report Form: Types of Chemical Reactions

I. *Combination Reactions:* Data Table. (These reactions will be demonstrated by the *Instructor*)

Rxn No.	Equations for the Reactions	Observations (Evidence for the Reactions)
1.	Magnesium (s) + Oxygen (g) → Magnesium oxide (s)	
2.	Calcium (s) + Water (l) → Calcium hydroxide (aq) + Hydrogen (g)	

II. *Decomposition Reactions:* Data Table. (These reactions will be demonstrated by the *Instructor*)

Rxn No.	Equations for the Reactions	Observations (Evidence for the Reactions)
1.	Sodium bicarbonate (s) → Sodium carbonate (s) + Water (l) + Carbon dioxide (g)	
2.	Potassium chlorate (s) → Potassium chloride (s) + Oxygen (g)	

Bronx Community College/CUNY
Department of Chemistry & Chemical Technology

Report Form: Types of Chemical Reactions

III. *Single Displacement Reactions:* Data Table. *(These reactions will be carried out by the students)*

Note: *Use approximately 2mL of aqueous solutions for testing*

Rxn No.	Equations for the Reactions	Observations (Evidence for the Reactions)
1.	Copper (s) + Silver nitrate (aq) → Copper(II) nitrate (aq) + Silver (s)	
2.	Zinc (s) + Hydrochloric acid (aq) → Zinc chloride (aq) + Hydrogen (g)	

IV. *Double Displacement Reactions:* Data Table. *(These reactions will be carried out by the students)*

Note: *Use approximately 2mL of each aqueous solution for testing*

Rxn No.	Equations for the Reactions (All Reagents are Aqueous Solutions)	Observations (Evidence for the Reactions)
1.	Sodium chloride + Silver nitrate →	
2.	Lead(II) nitrate + Sodium iodide →	

Bronx Community College/CUNY
Department of Chemistry & Chemical Technology

*Report Form: **Types of Chemical Reactions***

Rxn No.	**Equations for the Reactions (All Reagents are Aqueous Solutions)**	**Observations (Evidence for the Reactions)**
3.	Barium chloride + Sodium sulfate →	
4.	Sodium carbonate + Hydrochloric acid →	
5.	Potassium chromate + Lead(II) nitrate →	
6.	Sodium carbonate + Calcium chloride →	
7.	Iron(III) chloride + Ammonium hydroxide →	
8.	Sodium hydroxide + Hydrochloric acid →	